WHAT'S THE BIG IDEA?
Forming Hypotheses

Barbara A. Somervill

The Rosen Publishing Group's
PowerKids Press™
New York

For Lilly

Published in 2007 by The Rosen Publishing Group, Inc.
29 East 21st Street, New York, NY 10010

First Edition

Editor: Joanne Randolph
Book Design: Elana Davidian

Photo Credits: Cover © Steve Wanke/IndexStock Imagery, Inc.; p. 4 © Rich Iwasaki/Getty Images; p. 7 © Flip De Nooyer/Foto Natura/Minden Pictures; p. 8 © Art Wolfe/Getty Images; p. 11 © Bettmann/Corbis; p. 12 © Frans Lanting/Minden Pictures; p. 15 © Bonnie Kamin/IndexStock Imagery, Inc.; p. 16 © Richard T. Nowitz/Photo Researchers, Inc.; p. 19 © Heidi & Hans-Jurgen Koch/Minden Pictures; p. 20 Cindy Reiman.

Library of Congress Cataloging-in-Publication Data

Somervill, Barbara A.
   What's the big idea? : forming hypotheses / Barbara A. Somervill.— 1st ed.
      p. cm. — (Think like a scientist)
   Includes bibliographical references and index.
   ISBN 1-4042-3482-9 (lib. bdg.) — ISBN 1-4042-2191-3 (pbk.)
   1. Science—Methodology—Juvenile literature. 2. Hypothesis—Juvenile literature. I. Title. II. Series.
   Q175.2.S668 2007
   501—dc22
                                    2005029494

Manufactured in the United States of America

Contents

Scientists are curious people. They like to find out why and how things happen in our world.

The Scientific Method

Science is a key that unlocks the mysteries of our world. Good scientists follow a system to solve those mysteries. The system is called the **scientific** method.

The scientific method has several steps. Every good scientist follows these steps in order. You should start by asking a question. Next you will make a guess, or **hypothesis**. Test your hypothesis and record your findings. Finally you should use those findings to answer the question with which you started.

To be a scientist, start using the scientific method. Ask questions about the world. Plan experiments. Become a science **detective**.

The Problem or Question

At breakfast you drop an egg and it breaks. What a mess! You wonder why eggs break so easily. Then you think about eggs in a bird's nest. Why do eggs not break when birds sit on them? Are eggshells weak or strong? You want an answer to these questions. A hypothesis is the big idea behind your experiment.

You observed something that happens in the world around you. That is the first step. Then you asked several questions. Now narrow those questions down to one that you will answer by doing an experiment.

To follow the scientific method, write down exactly what you want to know. You do not need to use big words. Just be sure to write your question clearly. That question will guide your experiment.

Many birds lay eggs in a nest and then sit on the eggs to keep them warm. Eggshells need to be strong enough to allow the mother bird to sit on eggs without breaking them.

If you did research on eggs, you would find that they come in many sizes, shapes, and colors. These robin eggs are light blue and small.

Do Some Research

To narrow down your hypothesis, you need to do some **research**. Books, magazines, and Internet articles offer **information**. If you look up "eggs," you will find out that all eggs have the same basic shape. You will also learn that bird eggs come in different sizes and colors. Ostrich eggs, for example, are larger than robin eggs.

Professionals can also provide good information. You might ask a librarian, a chicken farmer, or a science teacher about eggs. You might also ask an **engineer**. Engineers often figure out what makes something strong or weak.

Write down what you find out from your research. Keep track of where the information came from. You will need to know that later.

The Meaning of "Hypothesis"

Every science experiment starts with a hypothesis. A hypothesis is a good guess that must be tested and proven as fact.

The hypothesis answers the question that the scientist is trying to prove. In forming a hypothesis, the scientist **predicts** what the result of the experiment will be. It can take many experiments to prove a single hypothesis.

The inventor Thomas Edison came up with a hypothesis. While trying to improve the way **telegraphs** sent messages, he thought it might be possible to record and play back sounds. He worked on producing a model that could do so, but he failed many times. In 1877, he produced a working phonograph. He proved his hypothesis was true, but it took many tries.

Thomas Edison was born in 1847. He is shown here with his phonograph. A phonograph records and plays back sound. Thomas Edison invented or improved more than 1,000 things.

What is the same about all these eggs? All the eggs have the same basic shape. This might give you a clue as to which feature of an egg has an effect on its strength.

Making a Good Guess

Now we know that a hypothesis is a scientist's best guess about what will happen. Scientists base their guesses on **experience** and research. You should do the same thing.

It is hard to answer several questions at one time. Scientists decide which question is most important to answer. They narrow their search. That way they can **focus** on one problem rather than on many.

Now try to answer the question based on your experience and research. What gives strength to eggshells? Is it size, shape, or color? Based on what you know, egg size and color change from bird to bird. These things do not seem to have an effect on the strength or weakness of an egg. It must be an egg's shape that has to do with its strength.

Writing the Hypothesis

Scientists keep track of their work by writing everything down. You, too, must record your ideas. You have to write down your hypothesis. Fancy words are not necessary. A simple, clearly written hypothesis is best.

You may need to make small changes as you go. As you get deeper into your experiment, you learn more about your subject. The more information you have, the better you can direct your experiment. Writing down your hypothesis helps you focus your work.

"I believe that an egg's shape makes it strong," is a simple hypothesis. Now you have a basis for a science experiment. You have something that you can prove by doing experiments.

It is important to take notes when you do an experiment. You need to keep track of your results. You also need to be able to do the experiment again and get the same results.

These students are working on an experiment in the classroom. Each student is following the same steps. If everyone gets the same results, then they will know whether their hypothesis was right or wrong.

Testing the Hypothesis

Create an experiment to test your hypothesis. To do this make a step-by-step plan to prove what you believe. You can run an experiment, take a **survey**, or do more research to help prove your hypothesis.

You may need to do your experiment more than once to prove your hypothesis true. If you run the experiment only once, you have nothing with which to compare your results. Your results might have mistakes, but you will not know it. A good experiment can be done many times and will get the same result each time.

When planning an experiment, you must follow certain rules. You can change only one feature of the experiment at a time. You should keep track of each step you take.

Making Some Changes

As you run your experiments, you discover that you left out something. Your hypothesis does not give a full or a true solution, or answer. Do not change your hypothesis. Add an explanation in your report that tells what added information you discovered.

While doing your experiments, you discover that some birds' eggs do break when the mother bird sits on them. This is not normal, or natural. What causes this problem? In some cases food has an effect on the mother bird. If she eats something with poison in it, she may lay weak eggs. The shells may be too thin, or they may be badly shaped because of the poison.

Include this information in your report. Point out that your results depend on using normal, healthy eggs. Your hypothesis does not include unnatural conditions.

As you do your research on eggs, you find out that only normal, healthy eggs will hatch. Here a chicken has broken free of its shell. It takes a chick 21 days to hatch from its egg.

Here a student has dropped a book from 3 feet (1 m) above the table with the eggshells on it. All the shells have broken. The student will make changes in the experiment based on this result.

Uh-oh, You Were Wrong

Some experiments do not work. You may have the correct hypothesis, but your experiment does not produce the right results.

Your hypothesis states that the shape of the eggshells makes them strong. To test it you cut three shells in half and placed them dome-side up on a table. You dropped a book 3 feet (1 m) above the table, and the shells broke. What went wrong? Have you used too much weight? Would the mother bird drop down with so much force? Make changes to your experiment until you prove your hypothesis true.

Remember Thomas Edison. He tried many experiments that did not succeed although he had the correct hypothesis. He said, "I have not failed. I've just found 10,000 ways that won't work."

What Did You Learn?

A hypothesis is the idea that starts every experiment. To prove any hypothesis, you must become a detective. You need to collect clues, or facts, about your subject.

A good hypothesis solves a mystery. The answer provides useful knowledge that we can apply in a number of ways. Proving the strength of eggshells is just as important as finding a new rocket fuel. In fact many bridges have been built based on what scientists have learned from the shape of eggshells.

Once a hypothesis is proved true, it is no longer a guess. It is a **theory**. A theory that is proved true repeatedly becomes a **principle** or law. People accept the idea as truth. The aim of every scientist is to find the truth.

Glossary

detective (dih-TEK-tiv) A person who finds facts and finds out who has done a crime.

engineer (en-juh-NEER) Masters at planning engines, machines, roads, and bridges.

experience (ik-SPEER-ee-ents) Knowledge or skill gained by doing or seeing something.

focus (FOH-kis) Think about.

hypothesis (hy-PAH-theh-ses) A possible answer to a problem.

information (in-fer-MAY-shun) Knowledge or facts.

predicts (prih-DIKTS) Makes a guess based on facts or knowledge.

principle (PRIN-sih-pul) A basic truth, law, or belief.

professionals (pruh-FEH-shuh-nulz) People who are paid for what they do.

research (REE-serch) Careful study.

scientific (sy-uhn-TIH-fik) Having to do with the use or study of science.

survey (SUR-vay) A set of questions asked of a number of people to find out what most people think about something.

telegraphs (TEH-lih-grafs) Machines used to send messages using coded signs.

theory (THEE-uh-ree) An explanation based on observation, experimentation, and reasoning.

Index

Web Sites

Due to the changing nature of Internet links, PowerKids Press has developed an online list of Web sites related to the subject of this book. This site is updated regularly. Please use this link to access the list:
www.powerkidslinks.com/usi/formhypo/